ANGORA
WOOL RANCHING
AND GOALS IN RABBIT RAISING

By WILLIAM E. OTTO and HEDLEY B. BURDEN

14th Edition

"Angora Wool Ranching" was First Published in 1937
13th Edition Published in 1976
14th Edition 1999

* All figures and statistics have been used and reprinted from the 1951 edition.

"Goals in Rabbit Raising" was First Published in 1972
2nd Edition 1999

ISBN 0-9685064-1-0

Published by
DIAMOND FARM BOOK PUBLISHERS
R.R.#3 Brighton, Ontario, Canada K0K 1H0

In USA
Box 537, Bailey Settlement Rd. • Alexandria Bay, NY 13607

email: diamond@intranet.ca • www.diamondfarm.com
1-800-481-1353 • Fax: 1-800-305-5138
Tel: (613) 475-1771 • Fax: (613) 475-3748

A FEW VIEWS OF OTTO'S ANGORA RANCH

TABLE OF CONTENTS

CHAPTER   **PAGE**

HEDLEY B. BURDEN
**"Goals in
Rabbit Raising"**

WILLIAM E. OTTO
"Angora Wool Ranching"

The 1999 Edition of "Angora Wool Ranching" is being published in memory of William E. Otto and the strong impact he has had on Angora ranching in North America. Mr. Otto pioneered the Angora Industry, and was known throughout Canada and the United States for the quality of his "Fashion Plate Angoras." He provided the backbone of the Angora Industry. This book is considered the oldest publication of its kind in America, and for years has been described as the text book of Angora ranching.

The 1999 Edition of "Goals in Rabbit Raising" (Section 2 of this book) is being dedicated in memory of Hedley B. Burden who ran a very successful commercial rabbitry, raising both New Zealand Whites and Californians. His 10 goals in raising rabbits will be of great help to anyone contemplating raising meat rabbits as a profitable venture.

INTRODUCTION

For several years, previous to the publication of
the first edition of this book, many requests were received
for detailed information on how I ranch my Angoras.

Beginners wanted a reliable guide to follow, while many
already growing Angora wool, realized that they, through lack
of information on the subject, were not as successful as
they should be. Since the first edition was published,
ANGORA WOOL RANCHING readers have sent in a
steady flow of letters of appreciation. Departments
of Agriculture and Experimental Farms have recommended
it as a reliable source of information.

In submitting the information contained herein, may I say,
I am not an author. The following pages are written by a
breeder, from a breeder's standpoint. The instructions and
advice offered are based on my actual experience
over a period of several years, and I can assure you
that not only have they proven successful on my own ranch,
but they will prove valuable to you and should be of
much assistance in your work of growing Angora.

WILLIAM E. OTTO

Angora Wool Ranching

THE AUTHOR'S SON, BOB OTTO
HOLDING ONE OF OTTO'S
"FASHION PLATE" GRAND CHAMPIONS

CHAPTER I
The Angora Rabbit

The Angora Rabbit is one of nature's most beautiful creations. An Angora, in full coat, is indeed a **"Thing of Beauty."** It enjoys immense popularity the world over, and rightly so.

Unlike other animals, raised for fur and flesh, they grow coats of silky wool: the world's highest priced. The wool is shorn – no killing of the animal. I can assure you this is one of the most pleasing aspects of the work. These little animals are your associates in business. They know you, and there is much satisfaction in knowing it will not be necessary to kill them in order to bring profits to your business.

This alone, has attracted many people to the growing of Angora wool. This is especially so where women are concerned, and women play an important part in the Angora industry. In Great Britain, several titled ladies have been, or are enthusiastic Angora breeders. The late Lady Rachael Byng, developed a strain of Angoras known the world over. At one time, owing to a scarcity of green feed in her immediate neighbourhood, she found it necessary to move her stock and equipment to another part of England, and she chartered a special train for the purpose.

In America we have many women in the industry. As a rule they are very successful and, in some instances, their Angora money is the mainstay of the family. Personally, I know of some very interesting cases.

In practically every part of America, Angoras are producing extra money for their owners. The fact that Angora wool is a non-perishable product, it can be shipped by parcel post or express, to the buyers and makes it possible for anyone, regardless of location, to engage in this work. Payment for wool is mailed to the grower.

A new source of income has been opened to many, handicapped because of their location, and who probably would not have an opportunity to add to their income if it was not for their interest in Angoras.

The grower may be located in a city or in the backwoods. To illustrate, I would like to quote from a letter received by a firm of wool buyers. It read: "I am living alone in a shack in the woods and all I have to live on is the returns from my Angoras. I built a lean-to from scraps of boards, covered with tar paper and banked with sawdust...."

On another occasion one of the leading British buyers published the following in Fur & Feather: "We often wonder if the public, even the Angora enthusiasts, realize what a wonderful part Angoras are playing in helping over this time of tribulation. We have many wool suppliers in and outside the United Kingdom, who are living entirely on what they earn from their Angoras, while hundreds of others depend upon their income from wool to pay their rent, clothing bills etc."

The wool grows rapidly with well-bred stock. It is sheared at 8 to 12 week intervals. This operation continues for the life of the animal, which is, on the average, 5 to 7 years. The first shearing, when the young are 6 to 7 weeks old, more than pays for their keep up to that age. From then to maturity, approximately 8 months of age, varying with individual animals, wool production increases with the size of the animal, until, when full maturity is reached, the annual yield may reach 20 oz. I wish to make clear to the beginner, the average production is lower, being closer to 14 oz. However, progress has been made, and will continue to be made, through diligent breeding, in the amount produced by the Angora.

The increased production is being accomplished by breeders who line-breed their stock, specializing in the Angora for wool production.

The Angora Rabbit

ENGLISH ANGORA DOE
ONE OF OTTO'S CHAMPIONS

I am very often asked why there are not more people growing wool. Up to a few years ago, suitable foundation stock could only be purchased by importing from England or other countries. Such stock commanded very high prices. Insurance and transportation rates were also very high, and very few people could afford a start in Angoras. Another reason is, that until recent years, only a few people knew such an animal existed. Today, it is not necessary to import foundation stock. Canadian herds were founded on the

finest imported stock. This stock is now thoroughly climatized. Our vigorous climate, and abundance of wholesome feeds, and years of selective breeding, has placed our stock in an enviable position, equal to, and vastly superior to many.

The Angora is a clean and interesting little animal. You will become very much attached to them, and in return they will pay you well for your efforts.

ENGLISH ANGORA DOE & LITTER OWNED BY
DIAMOND J. ANGORA RABBITRY, BRIGHTON, ONTARIO, CANADA

CHAPTER II
Angora Wool and Its Uses

The French were the pioneers in the manufacturing of Angora wool into yarn. Their processes were closely guarded, and admittance refused to the factories.

The Englishmen, sensing the possibilities, soon accumulated sufficient information, through their experiments in spinning Angora, to necessitate the importation into Great Britain, of approximately $2,500,000.00 worth of Angora wool, over a four year period. While it may be true that the average person is familiar with Angora yarns and garments; few persons know the raw wool comes from the Angora Rabbit.

Not only are these yarns the softest, warmest, and the most beautiful, but they command, by far, the highest prices. Angora is truly: "The Royalty of Wool."

Should you have a doubt concerning the value of Angora, it might be of interest to you, to turn to the yarn section of any leading mail-order catalogue, or visit any specialty shoppe, selling exclusive garments, and compare the price Angora commands in competition with other lines. Test them for appearance, softness and durability, and you will readily understand why textile mills pay profitable prices to secure regular suppliers of Angora wool.

Apart from its beauty, tests show Angora to be 7 to 8 times warmer than other raw wools. Angora wool possesses therapeutic qualities. Prominent French physicians recommend Angora garments for patients suffering from asthma, bronchitis, rheumatism, and similar ailments. The most skeptical are convinced of the superior qualities of Angora almost at a glance. Each season brings us a grand array of smart things made from Angora: coats, dresses, hats, gloves, sweaters, and other lines too numerous to mention.

* All figures and statistics have been used and reprinted from the original publication in 1951.

CHAPTER III
Advice to Beginners

The old adage, "Experience is the best teacher, but the fees are high," holds true as far as the Angora industry is concerned. Mistakes are costly, and with such importance attached to the selection of foundation stock, the beginner will do well to heed the information given in this chapter.

A man wishing to buy good clothes, goes to a merchant who has a reputation for supplying good clothes. He most certainly would not go to any other class of merchant. However, should he decide to go out bargain-hunting, he would do so armed with practical knowledge of clothes values. His past experience in purchasing clothes would place him on guard, give him the ability to judge value.

The beginner with Angoras, placing his order for stock is, in all profitability, making his first transaction, with no previous experience to guide him, and may easily become victim of an unscrupulous advertiser.

The Angora, because of the value of the wool it produces, attracts a type of individual to the selling of stock, to secure money from unsuspecting novices. He usually deals in the cheapest of stock and leaves the unsuspecting purchaser "holding the bag."

Occasionally stock may be picked up from: "My entire stock of Angoras and hutches, cheap," advertiser. If there is a legitimate reason for an advertiser wishing to dispose of his stock, if it is of high quality, a profitable deal may be made, but without such a reason, why is this advertiser disposing of his stock? Has he been a victim of the type of seller previously mentioned, and endeavoring to pass them off on anyone who will buy them? If he cannot make money with his stock, why should the beginner expect to? If the stock is of the quality necessary to produce wool of the right quality, in paying quantities, it is unlikely they would be for sale.

There is the possibility that a good buy may be made. However, the risk of getting something of little or no value, is far too great for any beginner to take, unless he can have some party, who knows Angoras, uninterested in the sale of the animals, inspect them and recommend their purchase.

Another advertiser claims his stock is of some well-known strain. It may be true that he did, at some time, purchase stock from a well known strain, but through improper care or unsanitary housing, causing disease, the vitality is so often impaired that its purchase is a poor investment, at any price. In this case of the advertiser, he is who owns the animals of average quality. He purchases a pair of Angoras from a well-known strain, but with one purpose in mind; to be in a position to say he actually purchased stock from that strain, but his sales may be from his own stock of animals.

It is not my desire to detract from the honest advertiser. It is to advise you of the seller whose sole interest is to get all he can, regardless of the methods used.

Another advertiser says: "Pedigreed Woolers," at a few dollars a pair. Angoras of the quality necessary to establish a paying herd, are not available at such prices. Pedigrees mean nothing. A pedigree is simply a record of the ancestors. A pedigree, genuine or otherwise, is easily made. However, no breeder can make any process in his breeding without his pedigree or record of his animals breeding. Without this, he would be mating his animals in a hit and miss manner with disastrous results. When one takes into consideration, the revenue from the sale of wool annually from a good animal, without the increase in stock, it is apparent that such an advertiser is giving something away, but one usually gets just what he pays for. To save a few dollars, the difference between poor stock and stock of proven quality, is leaving the purchaser open to lose his money and time, as he must eventually start all over.

Another class of advertiser is the "buy-back" advertiser. You are offered stock, with a promise to purchase wool from this stock. About all that is necessary to "Make $3,000.00 or more a year, with our Angoras," is to order stock, ship the wool back and get rich. Such an outfit is more clever than you, when it comes to playing their game.

There is a great difference in strains of Angoras. Some grow nothing more than white hair; a practically worthless product. Look to the breeder who has spent years of work and study to develop his particular strain – to excel in wool production, in show points, or both. He has, for years, built his business on his ability as a breeder and on the satisfaction given to his customers. You may safely look to such a breeder for advice. Decide upon the strain you are going to use, then buy direct from the originator. Do not gamble on second-hand stock.

* All figures and statistics have been used and reprinted from the original publication in 1951.

"FASHION PLATE"
GRAND CHAMPIONS

CHAPTER IV
Spinning and Knitting

The majority of growers sell their wool in the form in which it comes off the animals. In Great Britain, a very profitable business has been developed in spinning the wool on small spinning wheels at home. Many of these spinners knit their yarns into attractive wearing apparel, while still others weave the yarns into fabrics.

The following, taken from a popular monthly magazine, tells in a few words what can be done and is being done. It reads, "In the small quarters of your own back yard, you can grow mother's coat, sister's knitted suit, and brother's sweater, with a pair of socks, left over, for dad. With two dozen Angoras, you can produce enough wool to supply the family with the finest knitted garments the market affords."

Clean and free Angora wool, as it comes off the animal, may without carding or special preparation, be hand-spun into beautiful yarns; quite easily too. These yarns take dyes beautifully. The spinning, knitting, and weaving of Angora, into attractive articles of wear, offers splendid opportunity. Such garments, if well made, command high prices. The price realized for an Angora garment, not only allows a splendid profit for the time used in making the garment, but a sufficiently large margin. The wool used may be valued at twice, or more, the price of the raw wool as sold on the open market. Those who enjoy spinning and knitting, can not only produce the finest wearing apparel for the family, but have the opportunity to develop a very profitable business at home.

These garments, when well made, are recognized as being superior to machine made. Well-displayed samples of your work will usually provide you plenty of work. This is especially so if you cater to a tourist trade of the better class.

Angora Wool Ranching

A BATTERY
OF FASHION
PLATE HUTCHES

OTTO'S BUNGALOW HUTCHES
OUTDOOR HOUSING DEVELOPED
ON THIS RANCH

CHAPTER V
General Care and Management

One of the first things to know is the proper way to handle an Angora. <u>Never</u> pick an animal up by the ears. Lift by taking hold of the skin over the shoulders, supporting the animal by placing your other hand under the hind quarters.

Do Angoras drink water? I cannot understand why some people should think a rabbit does not drink, yet some of the oldest fanciers do. Always see that your Angoras have plenty of good clean water. If, upon receiving stock from a distance, they seem very thirsty, give them a little at a time until the thirst is quenched.

For the health of your stock, sanitation is of the utmost importance. See that your animals are not kept under filthy conditions. Apart from the health of the animals, and the appearance of your rabbitry in general, cleanliness is necessary if you expect to realize the biggest returns from the sale of wool.

Keep feed and water dishes clean, scalding or disinfecting at least once a week. Use a small scrub brush, washing the dishes in water to which has been added a little disinfectant. Stock housed on wire floors does not come in contact with the refuse, unless it is allowed to accumulate until the space below the floor can hold no more. In warm weather, hutches should be cleaned once a week. A clean rabbitry is of much greater importance than you may realize. Remember, strangers will not tell you that you have a dirty rabbitry, but they may tell others. Give your stock a chance. You will find a very profitable by-word around rabbitry to be "cleanliness."

If you wish to mark an animal for identification, a temporary mark, usually lasting two or three weeks may be made in the ear with a wet indelible pencil. For a permanent mark, it will be necessary to tattoo, which is a simple operation. Use a tattoo pen, plier, or sharp pointed instrument to make the perforation in the ear. Rub tattoo ink into this and, when it heals, you will have a permanent mark.

Beginners are sometimes under the impression that it is difficult to determine the sex of an animal. This is not so, but the younger the animal, the more difficulty it presents. It is a simple matter after the young are a month old. Turn the animal up and examine the sexual organs by pressing slightly: Does show a distinct slit, while a Buck's show a protrusion. Examine a mature animal to become familiar with the appearance of the organs.

Surplus bucks can be castrated, so that they may be run together, without soiling of wool, as it is the case with uncastrated Bucks. For the beginner, it would be better to witness the actual castration by an experienced hand, than to perform the operation from instructions. It really requires two persons; one to hold the animal while the other does the work. One holds the animal, back down, on his lap. The testicles are swabbed with iodine. Holding the testicle in the fingers of the left hand, it is forced gently against the covering skin. An incision is made in the skin, not too deep that the testicle is pushed out, and another incision is made in the thin covering of the testicle. This allows it to slip out easily, and then the cords attached are severed. It is not necessary to stitch, since there is usually little bleeding. Bucks should be castrated when they are three to four months of age.

Water crocks will freeze hard in the colder climates. It will not harm the animals if they eat the ice, however, there will be some breakage, and the crocks should be replaced with tins, such as one pound tobacco or coffee tins. The ice may be easily removed by placing the tin in a pail of hot water, the heat penetrates the tin quickly, so the ice will loosen enough to be dumped out.

Keep the rabbitry floor swept and everything clean and neat about the place. You never know when a visitor may call, so be prepared.

CHAPTER VI
Breeding Methods

Under no circumstances should an animal, under nourished or diseased, be used for breeding. The Stud Buck is of all importance and should be selected with great care. One Buck to 6 or 7 Does is sufficient. However, much depends upon the stamina of the Buck in use. Bucks of proper size and wooling qualities, full of pep, are the ones to sire good stock. If possible, only use a Buck once or twice a week for service, and it is better to use a Buck regularly than spasmodically.

When mating a Doe, one service is sufficient, and more only depletes the Buck's vitality, resulting in weak litters. Watch the Buck for weight, and should he become too thin, use another Buck alternatively for awhile. When breaking in a young Buck, do not condemn him if he does not come up to expectations for his first few services.

The Breeding Doe should not be overly fat. Such a condition is the real reason for Does missing. Do not use small, undersized animals for breeding. Does should not be mated at less than 7 months of age. The young arrive 30 to 32 days from the date of mating.

When mating a Doe, take her to the Buck's hutch. Some mate readily; others are more backward. When you place her in the hutch, keep your left hand over her ears and lightly grasp her shoulders. Slip your right hand under her body and raise her to a straightened out position. The Buck will then serve the Doe, collapsing momentarily. If a Doe persists in refusing a certain Buck then try her next time with a different Buck. Always handle the animal gently, avoiding fuss or excitement. Testing is done on the third and tenth day after the original mating by trying the Doe again.

The Doe, after having been mated, requires no special attention, until about ten days before littering. She must, of course, be well fed and cared for. About ten days before kindling, place a nest box in her hutch and give her some dry, clean straw for nesting.

I am often asked whether Winter breeding is successful. Personally, I like these litters. However, they are harder to get, especially in colder climates, and more so with young Does. As a rule Winter litters are more vigorous. To carry out Winter breeding with any degree of success, you must provide warmth for the young at birth. We have a maternity ward adjoining our office. The desired temperature is maintained with an oil burner. It must be comfortable for the arrival of the litter and although while the litter is being nested by the Doe it can withstand a lot of cold, you should always remember that the young are uncovered to nurse. Should one fail to get back into the nest, it is possible it may be chilled or freeze to death before being found. As soon as the young are running around smartly, they are transferred to regular hutches which, of course, are not heated.

Winter is not the natural breeding season, and Does are more apt to miss at this time, yet by feeding well and providing warmth you can be very successful with Winter litters.

Occasionally a maiden Doe will fail to make a nest and the babies will be found crawling about, and sometimes on the floor of the hutch. In such a case, shape the nest box, shear a few locks of wool from the Doe, and put a little in the nest box. Place the babies on the wool as well as a little over them and do not disturb further.

Going back to the preparations for the arrival of the litter. For the nest box, any wooden box about 15 inches long, 12 inches wide and 15 inches high will do. Take one end out, leaving the bottom edge up to a height of 5 inches, to prevent the babies from crawling out.

After the box and straw has been placed in the hutch, leave the rest to the Doe. There is nothing more interesting than to watch her build her nest. She may build the nest at once or not until a short time before she is due to litter.

Do not allow strangers in, see that there is fresh water in the hutch and do not molest her. I know it is tempting for the beginner and even for the older breeder, to see what is in the nest after she has kindled. This is not advisable. However, if your curiosity gets the best of you, then glance over the edge of the nest. If every-

thing appears okay, leave it at that. You may not see the young, as they may be covered with the wool which the Doe has placed over them. You may even see the wool moving, which should prove to you that they are there and well.

Personally, I examine each litter a few hours after it arrives and at regular intervals. Experience will teach you just how much any particular Doe will stand. Some do not want you even to open the hutch door at this time, while others place much more confidence in you. A couple of days after the young arrive, it is good to examine the nest. Remove the Doe from the hutch, place her in a box or basket, so she cannot watch you. You must avoid excitement while doing this. Make as little fuss as possible, keeping in mind that those babies belong to her, and that she has all the motherly instinct of any animal. Lift the box out, and with a lead pencil, or small stick, part the wool to uncover the young. Should the litter be too big, over 6 or 7, the extras should be removed.

After your nest inspection has been completed, and the Doe is back in her hutch, give her a piece of carrot or anything of that nature, to divert her attention and take her mind off the ordeal.

Should, for any reason, a Doe lose her litter, mate her in a day or so to avoid milk trouble.

CHAPTER VII
The Housing of Angoras

I t is very important that Angoras are properly housed. However, this does not mean that a lot of money must be invested to provide such equipment.

Stock must always be kept clean and comfortable. When good equipment is built, it should be made to save time in caring for your stock. The hutch described here is easily built and may be used inside any suitable building; one free from rain and draughts. It also may be used outside if given protection from beating rains and the blistering rays of the Summer sun. In the colder climates, for the convenience of the attendant, inside housing is preferable.

The hutches may be placed around the inside walls, two or three tiers high, to conserve floor space. If space permits, they may be placed in aisles, so much the better. Where no suitable building is available, a narrow, 6 or 7 foot house of any desired length, may be constructed. The end and back walls are tight, while the front is open to admit air and sunshine.

Angoras properly housed are seldom subject to disease. The object in producing Angora wool is to make money. Pay attention to your housing and you will have much larger wool cheques. There are two very important features in constructing a hutch for Angoras: the hay rack and floors.

A new type of rack, which originated on our own ranch, has revolutionized the hutching of Angoras. After years of experimenting with different kinds of equipment, a satisfactory hutch was constructed. It has given entire satisfaction and is used extensively by ranchers throughout America. It is a fine appearing hutch and it is not necessary to open a door to feed, water, hay, or clean your stock.

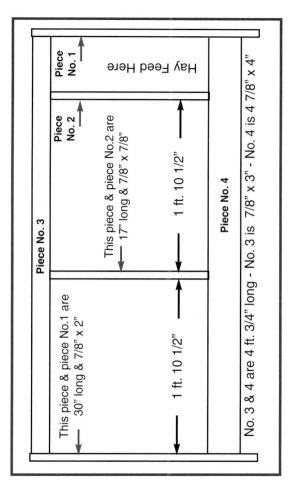

Piece No. 3

Piece No. 1

Hay Feed Here

Piece No. 2

This piece & piece No.2 are
17" long & 7/8" x 7/8"

1 ft. 10 1/2"

Piece No. 4

This piece & piece No.1 are
30" long & 7/8" x 2"

1 ft. 10 1/2"

No. 3 & 4 are 4 ft. 3/4" long - No. 3 is 7/8" x 3" - No. 4 is 4 7/8" x 4"

Angora Wool Ranching

By carefully following the information given herein, you will have no difficulty in constructing the "Fashion Plate" hutch. The first part to construct is the front frame, as shown on the proceeding page. The two large openings are for the doors.

The doors, shown on the next page, are made next.

Left Door -   Pieces No. 1 are of 7/8" x 2" lumber and pieces No. 2 are 7/8" square.

Right Door -   Pieces No. 1 are 7/8" square material and No. 2 are 7/8" x 2"

The framing around the feed dish opening, marked X, and the watering opening XX, is of 7/8" square material. Both doors, with the exception of the feed dish and water openings, are covered with 1" poultry netting. You will note that I do not give the outside measurements of the doors, as it is better to fit them to the frame openings. Cut a piece of galvanized iron slightly smaller than the opening XX, which is the opening through which the animals are watered, and at the upper end of this piece, drill two small holes. In the outside top edge of the door frame, screw in two small screw-eyes. Push the tin or flap in, so that when it closes, the bottom edge will be on the inside of the opening. It closes automatically.

The doors are hinged on the outer ends, so that one button will hold both doors, when closed. The back, ends, and top of the hutch, are of half inch lumber. The back is solid, and the same dimensions as the front frame. The ends, or width of the hutch allow a clearance of 1 ft., 8 1/2" from the inside of the front frame, to the inside of the back. You now have the four sides of the hutch.

The hay rack is made from the plan on the next page. This is of a different type to what is usually seen and it entirely eliminates any possibility of hay falling into the coats of the wool. All the hay eaten by the stock is pulled through the rack on a line with their faces.

Angora Wool Ranching

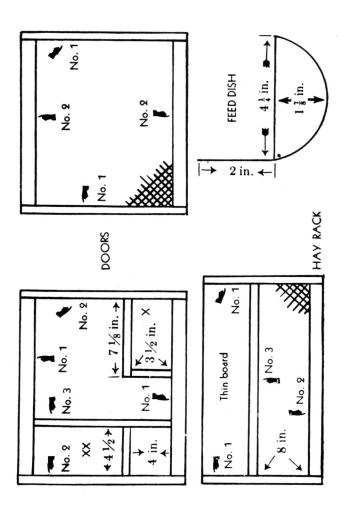

FEED DISH

$4\frac{1}{4}$ in.

$1\frac{7}{8}$ in.

2 in.

DOORS

No. 1

No. 2

No. 2

No. 1

HAY RACK

No. 2

No. 1

No. 3

No. 1

$7\frac{1}{8}$ in.

X

$3\frac{1}{2}$ in.

No. 1

No. 2

XX

$4\frac{1}{2}$

4 in.

No. 1

Thin board

No. 3

No. 2

No. 1

8 in.

The "V" type rack that is sometimes still used, is a very poor piece of equipment for wool ranchers.

Build the hay rack from the plan on the preceding page. Pieces No. 1 are 7/8" square, and 1ft. 9". Piece No.2 is 7/8" x 3" and 1ft. 6 3/4" long. Piece No. 3 is 7/8" x 2" and same length as piece No. 2 . The space between is covered with 1" poultry netting. Place the rack in the hutch, so that one end of it is flush with the post at the left of the hay rack opening in the front frame. After it is secured in place, fit a board in the bottom to prevent the hay from falling through to the top of the hutch directly below.

Any handy person, or tinsmith, can make the feed dish from the drawing. The dish is 6" long. Where the dot is shown in the drawing, drill a hole in each end of the dish. Then insert a stiff wire so the ends project 1 inch out of the end of the feed dish. These ends rest on two small, partly-opened, screw eyes placed in the right and left hand framing of the feed dish opening.

You are now ready for the floors. One feature of this hutch is that it may be converted from a wire floor to a board floor, suitable for a Doe and her litter, by substituting one set of floors for the other. To make the floors, cut 4 pieces of 7/8" x 2" material, 20" long. Make a frame out of these and cover with 5/8" square mesh hardware cloth. This wire is similar to that used for the screening of gravel etc. The floors rest on small blocks, nailed on the inside of the front frame and on the back wall. Make two such floors. Fit the space between the floors with a piece of lumber.

The hutch is now ready for the top, which is made of thin boards, tongue and groove preferred, and painted with a good black water-proof paint. You may have galvanized pans or trays made to put under the wire floors, or you may use a layer of sawdust, as a substitute for the pans, to absorb the moisture from the animals.

Modern, all-wire hutches, complete with automatic watering systems are now available from manufacturers.

CHAPTER VIII

Feeding Your Stock

The feeding of Angoras is a simple matter. The foods they consume, are easily produced in almost any locality. The feeding, as outlined in this chapter, has given excellent results on this ranch. Regularity in feeding is very important.

Our morning feed is grain. We feed 1 to 1 1/2 ounces per animal, of good clean oats – 2 parts oats and 1 part wheat (by weight) with a small portion of barley added, makes a fine grain ration. Crushed oats are excellent for a change. Fresh water is always in front of the stock. Hay, preferably second cut alfalfa, may be kept in the hay racks at all times. However, good results may be obtained by giving a good sized handful of hay for the evening meal. A piece of carrot, turnip, dandelions, or anything of this nature, is fed with the hay in the evening. That, in a few words, is the feeding schedule on this ranch. Mash may be most conveniently fed in pellet form. Pellets are mash in compressed form. They are fed just as they come from the manufacturers, no moistening being necessary, as with home mixed mashes. A very good formula for a home mixed mash is:

50 lbs. of Bran	25 lbs. of Middlings
25 lbs. of Oat Chop	15 lbs. of Fish Meal
1 lb. of Iodized Salt	1 3/4 lbs. of Linseed Meal

The amount may be increased or decreased, according to the size of your herd. This is moistened with water to a crumbly state, and fed at the evening feed. If you have milk instead of water, all the better, and especially so for Does with young or for growing stock.

If you have stock that seems lagging a little, try a moistened mash of two parts Bran (by weight) to one part Calf Meal.

Greens may be fed as gathered, however, **no wet or musty foods of any kind** should be fed under any circumstances. Dandelions, sow thistle, kale, lettuce, carrots, swede turnips, and other similar feeds are excellent supplements to the grain and hay rotations.

Greens should be fed sparingly to your stock, otherwise a bad case of scours may result.

The Doe provides nurse for her babies for the first weeks of their life, and it's important that she is well fed before and after bearing young. I give my Does all they will eat after littering and for a week previous to littering. An extra feed of bread and milk is excellent for a nursing Doe.

MODERN ALL METAL HUTCHES

When the young start coming out of the nest box, around 10 to 14 days of age, it is important that all food within their reach is clean and wholesome. My first feed for them is rolled or crushed oats, which is kept in front of them at all times. By the time they are 5 or 6 weeks old, they are on almost the same feed rotation as the older animals. Feed greens, carrots, etc. very sparingly to avoid diarrhea, or scours.

Grain builds weight, therefore, if an animal is too fleshy, reduce its allowance a little: if on the thin side, increase the allowance. A plentiful supply of greens increases wool yield. Do not overfeed your animals. I like them to be waiting anxiously for their feeding time. Should there be feed left from the previous meal, you are feeding too heavily, or the animal is off its feed. An undernourished animal is an easy victim for disease. Regularity in feeding, attention to the selection of clean, wholesome foods and an eye for condition, will keep your herd in the best of condition. All home-grown feeds cut down feeding costs, leaving better net returns from every wool cheque.

CHAPTER IX
Shearing the Angora

ANGORA "THE ROYALTY OF WOOL"

Before starting to shear, the coat of the Angora must be given a thorough brushing. With the proper hutching, there will be little hay, if any, in the coat. A wire haired brush with bristles set in rubber is the ideal brush for cleaning and straightening the wool.

Ordinary barber scissors are used for shearing. Have some 20 pound paper bags on hand, to hold the wool. Keep in mind, the less wool is handled after shearing, the better. A small surfaced table is used on which the animal is placed for shearing. Most animals seem to enjoy being sheared.

SHEARING: Part the wool down the centre of the back, brushing lightly to each side. Place the animal on the shearing stand, with its head to your left side, and by working from rump to neck, on the far side of the part, begin to shear. Do not get too much wool in the shears in an individual cut! As the wool comes off the animals, it is placed in the bags accordingly to grade. The different grades are explained in the next chapter.

Angora Wool Ranching

A BREEDING DOE WITH HER SHEARING OF WOOL

After shearing down the far side as far as possible, turn the animal so its head is toward your right side. Proceed to shear as before, and after the wool has been shorn from the back and sides, you proceed with the breast and belly coat.

The fingers of the left hand are placed under the chin, raising the head slightly, the four feet resting on the table, or stool. Shear across the breast, down to and between the front legs. Now raise the animal by grasping the skin over the shoulders, until it rests only on its hind feet. Now shear the belly.

Do not shear blindly! Make sure there is nothing in the shears but wool! A Doe might be damaged badly if her teats were cut.

The Doe uses wool for building and lining her nest as well as covering her young. Therefore, the breast and belly wool of a Doe that is to be mated in the immediate future, is <u>not</u> shorn.

If there is a mat, shear between the mat and the skin! There is always a line of free wool between mat and skin, but be careful of the skin.

In really cold weather, leave a little longer wool on the animal when shearing.

A little straw on wire floored hutches gives protection to stock shorn in cold weather. The Angora is hardy, but use your good judgment! Give your stock a chance!

You can purchase an electric clipper with special Angora blades.

CHAPTER X
Marketing Angora Wool

It cannot be too strongly emphasized that the finest of stock, housed in good hutches, may prove a poor investment, if you fail to market your wool in an intelligent manner.

Your failure to meet the requirements of the buyers of your wool will result in your wool being graded down, to such an extent that you are discouraged and leave the industry, disgruntled and an absolute failure so far as the production of Angora wool is concerned. There is no necessity for such a state of affairs if you will take the advice and follow the instructions of one who has had, at this time of writing, over 15 years of experience in the marketing of Angora wool. Your failure to follow these instructions will place the responsibility entirely on your own shoulders.

Remember, there is nothing difficult about the preparation of your wool. There are no processes requiring skill or anything necessary to make it difficult to secure the highest prices.

The only difference between the successful and unsuccessful grower, is that while both are given the same information, the same advice, the successful grower follows the information at hand, and enjoys the greater returns for his wool, while the other grower, not following instructions, complains. He becomes discouraged and eventually is forced out of business.

Keep in mind, buyers of your wool will not pay you for FIRST grade wool prices if your wool does not meet the requirements for that grade. Wool graders, employed by reliable buyers, grade your wool exactly as they find it. They will not stop to pick the particles of hay, etc., out of your wool, so that you may receive a larger cheque. It is your job to eliminate foreign matter from your wool, by brushing before shearing. You should aim to produce 85% to 90% first grade wool, and you can if you do your duty. Should you neglect to do so, then you must be satisfied with the returns received for your wool. The fault is yours, and yours alone.

Angora wool is a high-priced product. Textile mills cannot and will not pay such prices for wool which does not meet their standards. Wool may be of good length, the proper staple to grade No. 1, but it may contain specks, particles of hay, straw, etc., and because of this, it has to be classed as dirty wool at a much lower price.

Soiled or dirty wool must go through an additional process of manufacture which entails added costs. When you consider there is usually a spread of around $4.00 a pound between the two grades, you can easily realize the importance of giving the coat a thorough brushing before you start to shear.

Angoras of the proper quality, hutches with hay racks as described in this book; clean hutches, and a good wire-haired brush, are the prerequisites for top wool production.

Moth-eaten wool, or wool containing moth eggs, is worthless and usually destroyed as soon as received at the buyers. Do not leave the tops of the bags that hold the wool open after you have finished shearing. Fold the tops over and use an ordinary spring clothes pin for a temporary fastener. When holding wool in warm weather, place a few moth balls in each bag. Wool packed too tightly in the bags, or held too long, has a tendency to tangle, therefore, it is best to market wool as soon as possible after shearing.

Wool buyers have printed information giving their grades for the convenience and guidance of their shippers.

In a previous chapter, I referred to the advertiser who promises to buy back your wool if you purchase animals from them. This is an old racket, and while the shipper of animals will take the wool, he

* All figures and statistics have been used and reprinted from the original publication in 1951.

ANGORA APPAREL COURTESY GRAND'MERE KNITTING CO., LTD, GRAND'MERE, QUE., CANADA

Angora

KNITTED SPORTSWEAR

can, by short weighing and grading down, so sicken the beginner, that he looks for other markets.

In the meantime, the seller of the animals has his money for the animals.

Do not ship wool to any advertiser who has not been recommended to you by some breeder who can recommend a buyer. Fortunately such buyers are few, yet, during the course of a year, I receive numerous letters from growers, complaining. The following is a typical letter. It reads as follows:

"I would like to get the address of a reliable market. The only one I know of is at _____, but find their weights are very different from mine. Please note this: the first shipment was okay, but the second, they cut me down quite a bit."

That is their method of operation, and such a buyer should be avoided. There are thoroughly reliable markets of several years record of fair and honest dealings.

Angora Wool Ranching

THE MAIN BREEDING HOUSE AT OTTO'S ANGORA RANCH

CHAPTER XI
The Medicine Chest

Prevention of disease is better than treatment of disease. All livestock are susceptible to ailments. However, Angoras are hardy, and if properly taken care of, require little medical attention. Cleanliness is an important factor. Filth breeds disease.

An ailing animal should be removed at once from contact with other stock. It should be placed in a clean hutch by itself, and the hutch from which it was taken, thoroughly cleaned and disinfected.

The earlier the treatment is given, the quicker the animal will respond to treatment. It cannot be too strongly emphasized that disease is unnecessary. Its presence is an indication that you are neglecting your cleaning, or that your animals are undernourished. When you buy stock, make sure it is healthy before putting it into your rabbitry.

The best way to administer medicine is with a medicine dropper. Care should be taken that the animal does not bite the end off the dropper and swallow the glass. Put the end of the dropper in the side of the mouth, instead of directly in line with the front teeth, which are very sharp. Hold the animal down while giving treatment.

One of the first indications that the animal is ailing may be its failure to eat. Its eyes are dull and it acts sluggish. Experience will tell you just about what the trouble is, and what treatment, if any, is necessary. The eye of the Angora is a brilliant, sparkling ruby. The eye of an ailing animal is dull, pale, and lifeless.

Whatever the ailment, treatment should be regular, and must be continued until full recovery is made. You may have the occasional animal ailing, however, this is no need for alarm. The danger lies in your neglect to treat the ailing animal.

If, through neglect, your stock becomes diseased, don't fuss with them. You will save time and money by destroying the victims. Disinfect all equipment and make a new start.

Causes and Treatments of Common Ailments

WOOL BOUND The Angora grooms itself daily. It is a clean little animal and in its daily grooming, a certain amount of wool enters the stomach. If this wool does not pass through naturally, it will accumulate, and form a ball in the stomach, binding its contents and enlarging to such an extent that the digestive system becomes blocked, resulting in death. This condition is caused through improper feeding, lack of roughage – hay. Mash and hay prevent this, making it practically unknown with proper management.

The first indication of this ailment is that the animal refuses food. Give a teaspoon of Castor Oil or Syrup of figs, night and morning, until the digestive tract has been cleared. Do not give anything to eat until the animal appears to be recovering. Moistened mash is fine at this time.

SCOURS may be caused from overfeeding greens or frozen roots. It is usually fatal to young animals. Older animals usually respond to a teaspoon of Castor or Mineral Oil. Feed greens sparingly to young animals.

SORE OR WATERY EYES may be caused by a piece of wool in the eye, or by a piece of hay or straw. Bathe the eyes with Boracic Acid. A drop of Murine gives beneficial results too.

COLDS may be caused by excessive drafts, low vitality or allowing the animal to come in contact with its own manure. The nose is wet and the animal sneezes. Colds are contagious. Some are obstinate in responding to treatment, while others clear up in a few days. At the first sign of a cold, administer treatment, and continue until symptoms disappear. There are several good remedies.

Vicks VapoRub, rubbed on the nose at night and in the morning, also on the inside of their front feet, which they use as a handkerchief, is good treatment. Another treatment is to place 3 or 4 drops of any good commercial nose drops in each nostril in the morning and at night. A good home mixture may be made by mixing equal parts of Eucalyptus and Sweet Oil. In addition to either of the above treatments, a very effective remedy is 3 drops of Tincture of Aconite, in half a teaspoon of water. For young animals, use only 1 drop. Animals under 6 weeks of age should not be given Aconite. Give Aconite morning and night.

SLOBBERS This is wetness about the mouth. It is caused by some digestive disorder. A little ordinary Baking Soda, mixed in water, quickly rectifies this condition.

PREVENTIONS

• Cod Liver Oil and Iodine may be used to advantage at times.

• 4 drops of Iodine to the gallon of drinking water makes a splendid tonic.

• 2 or 3 drops of Cod Liver Oil twice a week is a fine builder for a backward animal, or for an animal recovering from an illness. A convenient way to feed oil is to put it on a crust of bread.

• Salt bricks, placed one to a hutch, may be made by sawing up into small squares, an ordinary cattle lick, either plain or iodized.

* For further information on preventions & remedies for Angoras, refer to "Otto's Angora Questions & Answers" available in most book stores or direct from the publisher.

CHAPTER XII
Making A Living From Angoras

With changing business conditions, a great many people who have never known what it means to be unemployed are suddenly brought to realize that they are but a small cog in the great wheel of business, and that their services may easily be dispensed with, or their salaries reduced.

Such conditions ruin many, while they benefit others. When brought face to face with the problem of making a living, many find they possess qualities they did not know existed within them; qualities developed through necessity.

Angora Wool Ranching
A BATTERY OF FASHION PLATE HUTCHES

Every ambitious person, although employed, should make some effort to develop, during their spare time, something which offers the possibility of expanding to such an extent that there need be no fear of unemployment.

I receive many letters from people asking if they can make a living from Angoras. This is a question that no person can answer intelligently. Such a letter usually comes from a total stranger and I have no way of knowing whether the person is energetic or otherwise.

You may give two different parties identical information on Angora Ranching, and one will make more money with 100 Angoras than the other will make with 500. It is not what one knows that makes him successful, but what he uses of what he knows.

According to your means, start to develop your business along sound lines, expanding as you feel fit. Good quality Angoras properly handled, provide an excellent income. I know an Ontario woman, working under the greatest handicaps, who earned over $2000.00 from her 200 Angoras in one year.*

Personally, I have never regretted making Angoras my life work. For years I held responsible positions which necessitated extensive travel and after many years of this, I much prefer working with my stock. My business increases each year and any ideas I conceive are used in my own business, instead of being handed over to an employer.

The growing of Angora wool is fast becoming an important source of income for many. There is an excellent opportunity for any ambitious man or woman who will go about it in the right way.

* All figures and statistics have been used and reprinted from the original publication in 1951.

CHAPTER XIII
Forms Used in the Rabbitry

R ecords must be kept of all animals, their breeding, and the performance of the animals in the breeding pens. The Brood Doe card gives an accurate record of what each Doe is doing, date she was mated, name of Buck to which she was mated, testing date, and the sex of her young.

The Stud Buck card shows what results you are getting from any particular animal. Should some produce better than others, those are the Bucks to use.

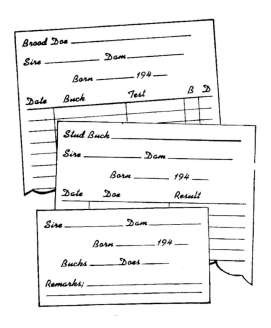

The other card is for growing stock, or may be used for stock not being used in the breeding. A Stud Buck or Brood Doe card is assigned to that animal and affixed to its hutch.

Wool records of the production of certain or all animals may be kept.

Pedigrees are supplied with animals for foundation purposes. The pedigree, shown on this page, is the special form we use on this ranch.

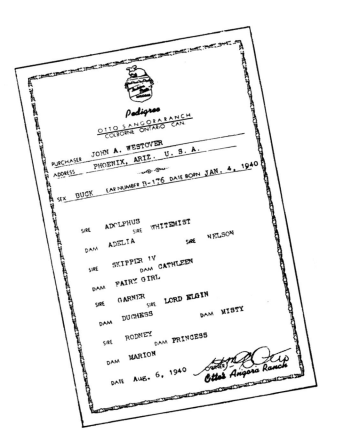

The Opportunity
This Industry Offers

Many families throughout America would welcome the opportunity to increase their income, if they could only find a way to do so.

To these men and women, the growing of Angora wool offers a clean and interesting means of making extra money.

The wool you grow may be turned into **cash** by selling to those who buy Angora wool or you may produce at home, hand-spun and hand knitted articles of the smartest wearing apparel. You can establish a **complete industry** right in the seclusion of your own home, realizing exceptional profits.

Angora wool is produced on small and large ranches throughout Canada and the United States. Growing this wool provides extra money for many families in all parts of North America.

Where you live is of little importance when it comes to the growing of Angora wool.

Turn to the yarn section of any leading mail-order catalogue and note the price of **Angora Yarns** in comparison with the other wools.

CHAPTER XIV
In Conclusion

Now, that you have finished this section, I sincerely hope the information it contains will help you. You have the necessary information to establish a successful Angora business. The methods you have learned of are in daily use by successful growers. To what extent you will profit depends upon how much of this information you apply to your work.

The fact that you have this book in your hands is sufficient proof that you are anxious to acquire information. Now that you have it, make use of it.

Those who pioneered the Angora industry had to grope along, experimenting and finding out for themselves, which was very slow and, incidentally, a very costly way of learning the business.

The pioneering is over. No person could have started Angoras under more discouraging circumstances than I did, because of the lack of reliable information. Friends and acquaintances ridiculed the idea that such an enterprise could be made to pay profits, let alone a living. Looking back twenty years or so, there is much satisfaction in knowing our hopes for the Angora industry were well-founded. Back in the early days, about the only reference you would find in trade circles concerning Angoras, was in connection with the sale of yarns. Today, Angora is featured everywhere. It is a favourite with smart dressers.

Section 2

Goals In

Rabbit Raising

By
HEDLEY B. BURDEN

INTRODUCTION

It is not unusual for the novice of Domestic Rabbit raising to be confronted with many questions relative to the possibilities of success, profit or expansion.

The time and effort that you put into the project, the manner in which you operate and the amount of capital invested will definitely have a bearing on these possibilities.

However, lack of experience, insufficient attention and lack of business knowledge are to be taken into consideration. Therefore, you should become as fully conversant as possible with all aspects of the rabbit raising business.

When you decide whether the keeping of rabbits is to serve as an enjoyable spare time hobby or as a profitable undertaking, you should use common-sense fundamentals which if put into practice are most likely to ensure success for your new venture.

The beginner should not be disillusioned into thinking that rabbit raising is a get-rich-quick proposition, however, meat rabbit production can and does pay well, provided it is carried out properly.

Raising rabbits for meat is becoming an important industry. It does provide unlimited and important possibilities no matter how small or how large a scale is decided on, whether it includes a few rabbits or several hundred.

If you are contemplating raising rabbits, this section is especially designed for you. It provides information for starting a rabbitry and goals to strive for, which you will find helpful.

Rabbit raising appeals to the youth as well as the adult for various reasons. It provides pleasure as well as profit. It stimulates activity. It is a pleasant pastime and a good retirement project.

The amount of space and the cost of equipment are at a minimum to start a small herd but to develop a large scale operation requires a sizeable capital.

Raising rabbits for profit is becoming big business. There are hundreds, even thousands, of people engaged in some phase of the rabbit industry and it is interesting to know that it is only in its infancy.

To raise rabbits successfully, continuous learning is required. Education will help people to adjust constantly to new devices and new ways of doing things. A rabbit producer cannot ignore what is going on so he would be well-advised to learn all he can about it.

HEDLEY B. BURDEN

TABLE OF CONTENTS

Goal #1
Raise Domestic Rabbits Desire

The first requisite for any venture is to have a desire to do. Without this fundamental principle, your ambition to raise domestic rabbits for either pleasure or profit will not materialize to any great extent.

It has been discovered that a large percentage of new participants to this fast-growing industry rapidly become discouraged and within a very short time find that it is not feasible to carry on with their anticipated high hopes of success.

Unless you have a genuine desire to participate in any worthwhile project, a successful outcome cannot be achieved. The raising of rabbits is no exception.

If you are really interested in establishing a profitable future business in any undertaking, a definite desire to accomplish this purpose will most assuredly crown your efforts with success to your entire satisfaction.

If you wish to raise rabbits you should consider seriously the purpose for which you want to enter this field. Your determination to make a success of your venture is very essential, and to accomplish this end you must possess, without reservation, a strong desire to fulfill all aspects of this worthwhile enterprise.

Plans

Before committing ourselves to anything, even when there is promise of great gain, it is well to stop to consider all the consequences.

It must be admitted that the changes brought about by research and new ideas have produced a host of problems. We cannot enter any new business lightly, expecting it to be as a good highway, free from bumps and pot-holes.

Any project is a problem that requires a solution whether it be large or small. It must be analyzed thoroughly in order to decide what steps are necessary to obtain the results desired.

So, to save time and energy and make possible the advantageous use of the resources at hand, planning the procedure is very necessary.

In the business world procedures are carefully scrutinized well in advance of work to be done. So must your plans be if you are to really enjoy a business in rabbit raising.

It is necessary to plan a step-by-step procedure before starting to raise rabbits – whether it be for a hobby, for show, for a supplement to your income, or for full-time employment.

Such thoughtful planning is essential to ensure wise choice of breed or breeds, proper material to use for the comfort of the animals, the use of proper tools for the tasks ahead and to give guidance for orderly work. There should be no haphazard methods applied.

In this technical age it is absurd to even think of the back-yard rabbitry as a means of developing a worthwhile business. We must use all the newer and conventional types of equipment at our disposal. A mistake or incorrect move might prove costly and discouraging to the beginner.

The mechanic, the banker, the dentist, the doctor, the carpenter, the teacher, in fact anyone who really succeeds in any undertaking, or anyone who creates, must plan an orderly procedure for everything he undertakes.

Building and operating a rabbitry is a good occupation and therefore requires things to be done by you in the right way to be capable, self-reliant and a credit to the industry.

Hobby

One conclusive proof that raising rabbits as a hobby is interesting and delightful is the fact that thousands of people are engaged in this fascinating work in Canada, the United States and other countries of the world. Rabbits are raised by children, 4-H Clubs, Boy Scouts, Girl Guides and many other youth organizations.

Many newcomers to the industry, including people in retirement, are becoming more involved and finding that raising rabbits as a hobby is a pleasant, relaxing and enjoyable pastime. Very little room is required and a few rabbits can be kept, for the hobby purpose, in a minimum of space.

Keeping rabbits as a hobby does not only lend itself to entertainment and other delightful pleasures, it often develops into a profitable one.

Without any experience whatsoever, people are becoming aware of the ease and economy with which the hobby of raising rabbits can be established, developed, maintained and enjoyed.

Sometimes we are surprised to see how well our hobby fits into our life pattern and proves very satisfying as a spare-time endeavor.

Show

Apart from giving great pleasure and, indeed, in many cases, excitement to the exhibitor, raising rabbits for shows has been a means of comparing animals and discussing many problems with satisfying solutions to these problems.

If it were not for shows – whether they be on a local, district or national basis – the rabbit would lose much of its identity. Therefore the showing of rabbits cannot be overlooked or taken lightly, for it is through this media that specialty clubs have been born, breeds have been promoted, and standards of perfection have been improved.

Generally speaking, shows are organized by Clubs, not only for the purpose of awards for the top-ranking animals, but they lend a fine opportunity for the novice to see for himself, at ease, the display of all breeds of rabbits and give him an opportunity to get acquainted with both breeder and breeds.

Meat for Family Use

Raising rabbits for meat seems to take preference, although there are considerable sales for the other purposes for which rabbits are produced.

During the years of World War I and World War II, when meat was difficult to obtain or rationed, the domestic rabbit was quickly recognized for its worth.

During the years prior to the wars, rabbit meat was a favourite item on the daily menus in the homes, hotels, and restaurants, as well as during the war years.

Today the public is becoming more aware of the high percentage of proteins contained in the white tender meat of rabbits, as well as being very digestible and second to none in food value.

Whether you live in the city or a rural area, raising a few rabbits for family use is simple and practical. Meat is an essential part of our diet but with costs so high, many people cannot afford to buy the better cuts. Therefore raising rabbit meat for home consumption is a good and satisfying proposition for those wishing to do so.

Supplement Income

The idea of raising rabbit meat to supplement one's income is becoming very interesting for many people. The extent to which this will help may be determined by the time and effort you devote to this work.

Many people raise rabbits as a part-time work and by doing so supplement their income and derive a great deal of pleasure at the same time. It is gratifying to people in the low income bracket to find that their earnings are being supplemented from the income that is derived from the sale of the rabbit raised.

Thousands of rabbits are raised and sold for this purpose and often develop into gratifying results. It is surely an excellent project offering both pleasure and profit, for spare time can be turned into cash.

When Edward H. Stahl purchased five rabbits for $15.00 little did he realize that his spare time venture would develop into a $350,000.00 a year business.

The possibilities are unlimited in the rabbit raising undertaking and it is surprising to see what a small beginning can eventually become.

Full-Time Employment

Raising rabbits on a full time basis can very well mean the difference between success and failure. To you who are thinking about the commercial aspect of the business, with little or no previous experience, the accompanying information should be of great help.

Today, rabbit meat, as a commercial product, is certainly being produced in large quantities, nevertheless it is a fact that the demand far exceeds the supply.

Since rabbits are being kept and processed today under much more sanitary conditions than was the case years ago, and since this pertinent fact has been impressed on the general public, the sale of rabbit meat has shown a very substantial increase.

The possibility of being involved in raising rabbits for commercial purposes was never so rewarding as it is today, providing of course that you strictly adhere to essential factors such as will be found in the pages of this book and in other publications.

A good reason why people raise rabbits on a full time basis is given in, "Rabbits for Profit and Pleasure" quote, "The reason why others raise rabbits as a full-time operation is they want to enjoy the pleasure, satisfaction, independence, and profit. It is an adventure into the joy of abundant living as he pursues the business intelligently and methodically."

Goal #2
Establish a Market

Assuming that you have decided to enter the field of commercial rabbit raising as a meat producer, it would seem logical that the next important approach is to establish a market for your output.

It would spell failure, even though you were successful in producing any commodity, if there were not a ready market available for that product.

There are many ways to establish a market for rabbits, giving the raiser an opportunity to add greatly to his income. For instance, he could sell rabbits for other purposes besides meat – such as laboratory stock, breeding stock, wool, fur, pets, and even fertilizer.

It is recommended that a consistent, year-round market be secured. If you do not secure a good, substantial market for your product before you begin, it could cause you some concern later on.

As stated previously, the demand for domestic rabbit meat is greater than the supply, therefore there should be no problem to establish a market.

It is absolutely necessary to be able to sell the production at a profit in order to be a success in the rabbit industry.

If the beginner lives close to a packer of rabbit meat he is in a very fortunate position, providing the packer offers a year-round contract. Those who buy rabbits only on an occasional basis are more of a hindrance than a help to the breeder as well as the industry.

Before making any commitment whatsoever, you should thoroughly investigate all market possibilities. If you live in an area remote from any packing plant, try to create your own market for ease and convenience. In some areas markets for rabbit meat are well-established at stores and meat shops.

Markets for your rabbits can be established by many means and here are a few suggestions in this direction.

Rabbit meat is sold to local outlets such as butcher shops, hotels, hospitals, and to individual families including relatives, friends and neighbours depending, of course, on the size of your rabbitry or the quantity of your output. So, by personal contact with these outlets, a market for your rabbits could be found and, indeed, help to build up a sizable business.

Thousands of tons of domestic rabbit meat are sold to markets which have been established to take care of this community. So, if your business develops to such an extent that you become a large producer then by all means you should contact these establishments and secure a market for your entire output. This will be necessary in order that you may enjoy a steady profitable income.

Occasionally rabbit raisers find their geographical position a problem to make personal contact with prospective buyers. In a situation of this nature it is advisable to make contact through the use of the mail in order to secure a market.

Advertising in local papers, show catalogues and hand bills is also a means of finding a satisfactory market not only for meat but for anything saleable from your rabbitry as well.

Again, a fair market could be found for quantities of rabbit meat by posting signs on bulletin boards in local stores and meat markets. Putting on displays, the hanging of your shingle, and having a roadside market, are also means of creating interest and eventually selling your product.

Goal #3
Use Modern Equipment Cages

Ith the assumption at this point that you have made up your mind to enter the meat rabbit production phase, we will now proceed to use the equipment that will be necessary for this purpose.

While many types and arrangements of cages have been suggested and found satisfactory to many rabbit-keepers, the author (Hedley Burden), who has had several years experience in operating a successful rabbitry will outline for your consideration and help the type of cages he has found very comparable and adequate.

There is a satisfactory as well as an unsatisfactory type of cage construction. The use of the all-wire cage is definitely advisable if you really want to operate a successful and profitable rabbitry. The wire cage is easy to make by the handyman or, if you prefer, they can be purchased from equipment dealers.

Wire is recommended for several reasons. They are easy to clean, making them hygienic, and the droppings fall to the pits below where they can be cleared away. If all-metal cages are used, diseases can be prevented and controlled.

This type of cage is being used more and more in modern installations. The cost is greater than the wood-wire type but they are more permanent and require so little or no repairs that in the long run are found to be less expensive.

Someone has suggested that one of the essential things in equipping a rabbitry is to keep the cost down. While we agree that inexpensive materials have been used and good rabbits have been produced it is a proven fact that the all-metal cage is by far superior to any other type.

You should build with the best modern material available if you want to obtain the fullest development of your rabbit project or business.

The arrangements of the cages in your rabbitry whether in one, two or three tier will largely be determined by the space available and the convenience the owner wants.

If plenty of space is available the one-tier arrangement is the most satisfactory because it saves the attendant from stretching or stooping when feeding and caring for the animals. Another convenience, the attendant can observe the rabbits better with the one-tier set-up.

Of course the two-tier as well as the three-tier have advantages too, in that they allow for more cages under one roof. But, if you really want convenience in cleaning, feeding, and observing the rabbits, the one-tier is considered the best.

The size of the cages will depend on the size and weight of the rabbit or rabbits it is meant to accommodate.

It is recommended that the floor space be large enough to allow about one square foot for each pound of mature, live weight of the rabbit occupying it, but it has been found that a cage with eight square feet is quite adequate for an adult rabbit of medium breed to raise her young.

It is not necessary to have the floor sloping if the cage has a wire mesh floor. Part wood and part wire mesh or slat wood is not required and in modern commercial rabbitries the solid floor is a thing of the past.

Before giving details of a suitable cage arrangement the following wire and gauge sizes are recommended.

Sides –	1" x 1" x 15",	16 gauge
Top –	1" x 1" x 24",	16 gauge
Bottoms –	1/2" x 1" x 24",	14 gauge
Partitions –	1" x 1" x 24",	16 gauge
Doors –	1" x 1" x 15",	14 gauge

Wire mesh is galvanized before or after welding; the latter is recommended because it lasts longer without rusting.

The size of a single cage for the medium breeds should be a minimum of 15" high, 24" deep and 40" long. (Fig.1) This type of cage construction should be so positioned in a rabbitry that the bottom of the cage is at least 28 inches above the ground to facilitate cleaning underneath and to keep drafts from the rabbits.

This is suitable for a one-tier arrangement. For a two or three tier plan, the minimum height for the floor should be at least 16 inches.

Cages may be hung from the ceiling by means of chain or wire.

FIG. 1
A Single Cage Unit

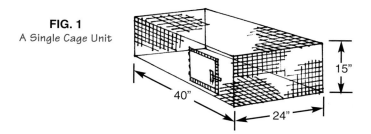

The double cage unit (Fig. 2), known as the Five Star Style is becoming very popular for ease and convenience. It has been proven to be quite satisfactory and is recommended.

The size is twice the length of the single cage, (i.e.) 80 inches. The width and depth will remain the same.

With this arrangement a space of 6 inches may be included for the purpose of a hay manger in the centre to serve the rabbits in each department if the owner wishes to feed hay in addition or as a supplement to the regular feed.

The self feeder is attached to one of the doors while the other door is used for placing the nest box and removing it. By placing the feeder on the door it can be easily checked or cleaned when necessary without removing it. By having the nest box in this section it can be removed partially to facilitate inspection of the litter while the doe is completely prevented from jumping in and out of the box while the inspection is in progress.

The nest box will slide on two runners which will be removed at the same time the box is, usually after weaning. The size of the runners is to be 1" x 1" x 24" made of dressed pine, cedar, spruce or fir. The purpose of the runners is to make it easier to remove and prevent the box from catching in the wire. When the nest box is removed, extra space is available for the young to grow and develop.

FIG. 2
One Section of the Five Star Double Cage

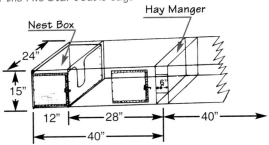

To make a cage (Fig.2) cut the wire to the desired sizes and fasten the pieces together with flat or J clips proceeding as follows:

1) Cut one piece 1" x 1" x 15", 208" long, fold at 80", 24", 80" and join to form two ends and two sides.

2) Cut one piece 1/2" x 1" x 24", 80" long and fasten to No. 1 for the bottom.

3) Cut four pieces 1" x 1" x 15", 24" long and fasten at the 12", 37" 43", 37", and 12" graduations respectively for partitions. Cut out openings in two of these dividers to match with the nest box openings.

4) Cut one piece 1" x 1" x 24", 80" long and fasten for top.

5) Cut four door openings to finish size 12" x 12". When cutting the door openings it is advisable to allow one inch on the top, bottom and latch side to be folded back to prevent injury to the animal and the owner.

6) Cut an opening to finish 6" x 10" for the hay manger. Keep this opening 3" from the bottom and 2" from the top of the cage.

7) Cut four pieces for doors size 1" x 14" x 15" using 14 gauge wire. Fold over 1" on the top, bottom and side (Fig. 3). This makes the door stiffer and prevents the hands from becoming scratched while attending to the animals. The finished doors will measure 13" x 13". Fasten the single side using J clips for hinges.

FIG. 3

Steps for Door Construction

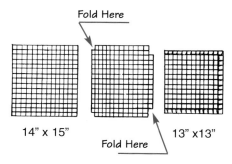

Fold Here

14" x 15"   Fold Here   13" x13"

Nest Box

The nest box is a very important part of a rabbitry and therefore must be so constructed to facilitate comfort for the rabbits and ease of handling and cleaning by the attendant.

The most common sizes of nest boxes in commercial rabbitries measure 16 to 20 inches long, 10 to 12 inches wide, and 8 to 10 inches high. The variations are given to accommodate the different types of commercial breeds.

The open nest box, (Fig.4) is quite satisfactory if used in a well-built rabbitry. For construction, use 1/2" fir plywood for the sides, the ends and the bottom. While rabbits will often chew wood, the amount of damage done to this type of box, after being used for several litters, has been very little.

A bill of material for the construction of an 8" x 10" x 18" common open nest box is printed below.

2 sides of 1/2" fir plywood 8" x 18"
2 ends of 1/2" fir plywood 8" x 9"
1 bottom 1/2" fir plywood 9" x 17"

When the stock has been detailed, assemble with 1 - 1/2" ardox nails. Cut a slot in the front end leaving four inches from the box and six inches wide to make it easy for the doe to enter and exit.

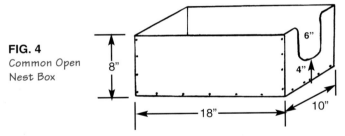

FIG. 4
Common Open Nest Box

8"

6"

4"

18"

10"

For the Five Star Style of cage (two in one) the nest box is somewhat different in that it is larger and the opening is on the side rather than the front. However, it has proved to be very satisfactory and worthy of recommendation.

The overall size of this type of nest box is 11" x 11" x 23". (Fig.5). To construct, proceed as follows:

1) Cut two pieces of plywood 1/2" X 11" x 23" — Sides

2) Cut two pieces plywood 1/2" x 10" x 11" — Ends

3) Cut one piece plywood 1/2" x 10" x 22" — Bottom

Two boxes will be necessary for each double unit cage, one for the right side and one for the left side. The openings will be opposite each other when placed in position. The assembly is the same as for the common nest box.

FIG. 5
Five Star Nest Box

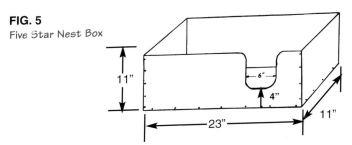

11"
6"
4"
23"
11"

Feeders

Feeders made of sturdy galvanized steel are recommended because they are long lasting, more sanitary than other types commonly used and they can be filled without opening the doors of the cages which makes it very convenient for the keeper. They are a real time and money saver.

They should be hung three or four inches above the cage floor to make it convenient for both adult and young rabbits to feed on the pellets contained in the feeder.

Contamination of feed is prevented and the rabbits can all get a fair share of clean fresh food. If these feeders are used in your rabbitry you will find that they are more satisfactory and in a short time will pay for themselves because of the time saved on labour costs. (Fig.6).

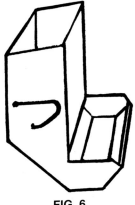

FIG. 6
Galvanized Feeder

Waterers

Because domestic rabbits are generally kept in wire cages and are not fed fresh green foods in abundance, as was the case in the past, they need a full supply of fresh, clean drinking water. Young rabbits drink more water as a rule, than adult rabbits. If water is supplied in crocks, cans or drinking bottles, these containers should be kept clean and filled several times a day.

This means plenty of work for the attendant. Consequently, the automatic watering system is by far superior to any other watering device because it supplies an abundance of fresh clean water to the herd.

This system can be easily installed by a handyman and the laborious and time-consuming task of filling other types of waterers, disinfecting and washing can be eliminated.

The young rabbits quickly learn to drink from the valves and, because a constant supply is readily available, over-crowding is eliminated at drinking time.

It is suggested that you use the automatic watering system right from the beginning if at all possible, regardless of the size of your rabbitry. If you are unable to do the installation yourself you could get practical help from the supply house who deal in this commodity.

The requirements for the automatic watering system include, 1/2" or 3/4" plastic hose, dewdrop valves, a breaker tank or a water pressure reducing valve and special tools for boring the holes and fitting the valves.

The hose is placed on the outside of the cages, preferably on the back, about 8" from the floor of the cage with a valve for each, installed at a 45 degree angle. Full instructions will be found in the suppliers manual.

Tools and Equipment

Every keeper or owner of a commercial rabbitry should have at his disposal a number of small tools and equipment necessary for proper maintenance and carrying out of daily chores.

Records

In order to be successful and maintain good business practices you should definitely keep precise records.

It does not necessarily have to be an elaborate bookkeeping system but a simple method of keeping a record of every rabbit in your rabbitry should be maintained.

To aid with this work you should keep up-to-date cards to give information on breeding date, service date, date does kindle, number kindled, number lost, saved, weaned, etc.

ITEMS FOR THE RABBITRY

J-Clips

Sprayer

Whisk

Dewdrop Valve

Torch

Scale

Shovel

Manure Fork

Wheel Barrow

Rake

Knives

Clock

Here is a partial list:
- Scales
- Weighing Cage
- Feed Cart
- Wire Cutters
- "J" Clip Pliers

- Hog Ring Pliers
- Hammer
- Saws
- Screw Driver
- Measuring Tape
- Gas Torch

- Shovel
- Wheel Barrow
- Spray Gun
- Brooms
- Manure Fork
- First Aid Kit

Other tools should be added as your business grows

You should also keep records of special chores to be attended to – such as when does are due to kindle, when nest boxes are due, does to be palpated, litters to be weaned, and rebreeding to be done.

It is wise to keep a record of other items to be attended to – such as treatment of diseases, checking leaky valves, cleaning and disinfecting, and any other items that you should get done on time.

Cards that you should keep for these purposes are: Hutch cards, Stud cards, Daily Record cards, Daily Chore cards, Sick Log cards, and others. Feed companies will gladly supply these cards for your aid in keeping your rabbitry well run.

Records pertaining to cost of operation; listing various expenditures; those pertaining to management and maintenance, and items pertaining to infertility and mortality should be kept.

Finally, records should be kept relating to selection of future breeding stock, the efficiency of each breeder, including mating results, number of bucks and does in each litter, milk yield in does, weight of litter at given periods, and resistance to various common diseases.

HUTCH CARD

Name or No. _____   Born _____   Cage No. _____

Sire _____   Dam _____

Buck	Date	Tested	Kindled	Born	Died	Saved	Weight

STUD CARD

Name or No. _____   Born _____ Sire _____ Dam _____

Doe	Served	Kindled	Number	Saved	Weaned	Remarks

CHORE CARD

Date _____   Remarks _____

Does to be Bred	Nestboxes	Testing	Weaning	Misc.

Tattooing

Rabbits may be marked in various ways in which to establish a permanent or temporary identification. The system of permanent marking which merits the most consideration is tattooing. Tattooing is definitely satisfactory for show rabbits and is most desirable for pedigree and registration purposes.

You should endeavor to tattoo all breeding rabbits in your rabbitry for permanent identification. When weaning young rabbits it is advisable to mark them so that later they may be correctly identified. This will certainly effect breed control.

In order to do this work you should possess a good tattoo set. They can be purchased from supply dealers. Tattoo sets consist of letters and numbers which are tattooed in the inner surface of the rabbit's ear by using the plier-tong tattooing instrument to press the pinpointed characters; perforating the ear.

The inside of the ear should be cleaned with petrol, alcohol or some other preparation to remove any dirt or grease. When the tattoo mark is made, black or blue or sometimes red, india ink or special ink for this purpose, is massaged well into the tiny holes. Some prefer to apply the ink prior to the marking and follow the massage afterward. It should be remembered that the left ear is used for the pedigree purpose while the right ear is reserved for registration marking.

Tattoo Box

The adjustable tattoo box (Fig.7) is a great aid to one person in marking rabbits. The construction is relatively simple for the handyman and should be part of the equipment found in all modern rabbitries.

Fir plywood is recommended for the entire construction along with the simple, inexpensive hardware required. Use 1/2" plywood for the sides, ends, bottom and top, while 1/4" material will suffice for the dividers. The overall size measurers 12" wide, 10" high and 22" long.

The floor should be moveable so that it may be raised or lowered into slots spaced so to accommodate the size of the rabbit being tattooed. Several small boards should be made to slide in slots spaced 1" apart to be adjusted at the back of the rabbit to keep it confined to the front of the box.

The top, or cover, will contain an opening about 2" x 3" to allow the rabbit's ear to protrude in preparation for the tattoo job to be done. It is fastened to the box with common butt hinges size, 1-1/2" x 1-1/2" at the back end and held locked in position at the front with a clasp.

FIG. 7
Adjustable Tattoo Box

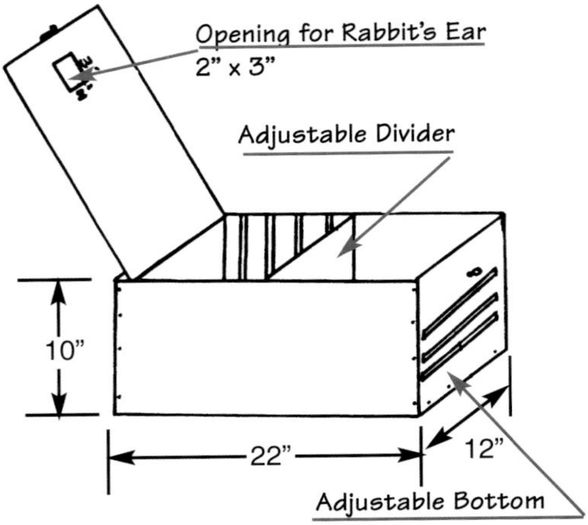

Opening for Rabbit's Ear
2" x 3"

Adjustable Divider

10"

22"

12"

Adjustable Bottom

Goal #4
Have Adequate Housing

There is a tremendous difference in the type of housing for rabbits and it will depend to a large extent on climate conditions, the location one wishes to establish, the size of the operation and the extent of capital available to be invested.

Before proceeding with the arrangements for housing the rabbits it will be necessary to decide on the type of shelter to be used and the phase of the industry with which you wish to become associated.

If you plan to raise rabbits as a hobby, or for meat for family use, or as a supplement to income, then the type of housing will differ from the arrangements that are necessary for a full-time operation on a commercial basis. The basic facts, however, should remain the same no matter for which purpose you raise rabbits.

Let us assume that you have decided on building a commercial rabbitry on a scale to house a sizable herd. When the location has been decided on and a check has been made concerning local by-laws or other legal restrictions that are likely to interfere with your project then, and only then, will it be feasible to proceed with the task of constructing the establishment.

We will confine ourselves to the equipping of an indoor rabbitry to be built where the climate conditions are such that a building is necessary to house the animals for most, if not all, of the year.

A rabbitry should be so planned that the construction and equipment will facilitate handling the rabbits with a minimum of time and labour. Construction should be simple, practical and efficient.

Make ample provision for adequate lighting and fresh air but one must guard against strong drafts, winds, and extremes of temperature.

A rabbitry should be so constructed that it will be efficient but at the same time within the limit of your financial budget. Good sanitation should be maintained by having a good drainage system.

Make the rabbitry wide enough to accommodate the number of cages you need for two, three, or more rows, and high enough to permit two or three tiers if you decide to go beyond the one-tier arrangement.

The alleys, at least 3 feet wide, constructed of either wood or concrete, are suitable, but open pits should take preference over full floors.

A well-constructed rabbitry will lend a good appearance, which is to be given to future expansion before the building is erected so that the owner will be spared headaches at some future date.

Arrangement should be made to take care of equipment and food and any supplies, including nest boxes and straw.

Ease in cleaning and convenience in feeding are important factors to be taken into consideration in planning and constructing a rabbitry.

Provision should be made to protect the rabbits from predators such as rats, weasels, snakes, and cats. This provision can be effective by having the footing for the concrete wall at least two feet in the ground.

The roof of the rabbitry should be built of material that provides insulation both against cold and heat. A wood roof covered with asphalt shingles is excellent but it costs more than heavy roofing felt.

The attendant should have comfort to perform his duties if a good job is to be done in the winter months as well as when it is naturally warmer.

Proper ventilation is very essential. If the rabbitry is not adequately ventilated, ammonia odours will be prevalent and condensation will take place. Ventilation should be controlled at all times in the rabbitry and this can be accomplished with ample outlets, louvers, high windows, piping in the walls and electric fans.

Heating the rabbitry is necessary for comfort for all concerned during the colder months. This can be done with oil space-heaters or electric heaters.

The temperature should be controlled and kept to about 50 ° F.

Rabbit Ranch Shed

It is not the purpose here to give detailed plans or step by step procedures in constructing a rabbitry, but rather to mention goals worth striving for.

Goal #5
Select a Good Breed

In order to ensure the success of your venture you should use clear-thinking fundamentals when purchasing your foundation stock.

You should contact responsible breeders of the variety you desire to raise if you have had little or no experience prior to this undertaking.

If you live in a community where rabbits are exhibited you should visit these shows and see for yourself the kind of animals that are on display. Failing to be satisfied in this direction, you should communicate with some rabbit association for information and contact Specialty Clubs.

You should become associated with a local club, if one exists in your area, for it is through this media that you can receive first class information on good foundation stock to start your rabbitry.

Again, if you have had no previous experience be sure you start on a small scale. All too many have been disappointed to find, after a short time, that the success they anticipated at the beginning did not materialize because too many rabbits were purchased to begin with.

First-class stock is certainly not cheap but it will pay in the long run if top quality rabbits are secured. You want foundation stock to be the best; rabbits that are vigorous, healthy, active, clean, and good producers.

The rabbits you purchase should have full bright eyes, slightly moist noses, clean erect ears, dry forepaws, well-shaped tails and feet, smooth and sleek fur, and in short, the best your money can buy.

After you have decided which phase of rabbit production you wish to specialize in, select the breed best suited to your purpose.

You should know that all breeds of domestic rabbits are suitable for meat, but the difference will be found in the purpose for which the meat is used. For instance, not all rabbits will have the same good percentage of meat dress-out, and therefore are not suitable to raise for commercial purposes.

Rabbits raised for fur or wool are usually recommended for meat production, such as the Angora, the Belgian Hare, the Beveren, the Rex, the English Spot and so on.

Those that are definitely recommended include the New Zealand, the Californians, the Champaigns and the Palominos to name a few.

In Canada and the USA, the New Zealand White and the Californian are the breeds commonly raised for meat. However, the Palominos, both Golden and Lynx, are proving to be a promising breed to raise for meat because of their high dress-out potential.

Although it is possible to raise two, or even three breeds of rabbits in the same rabbitry, strict attention must be given in order to keep them separated.

It is recommended that you do not cross-breed these rabbits because they become less and less productive and the vitality is lowered with each succeeding generation. A good suggestion is to start and concentrate on one breed.

It has been intimated that it is wise to start small. Some prefer to begin with a trio of registered, ready-to-breed stock, while others prefer to buy young rabbits and grow with them. It is a matter of choice and should be left to the discretion of the rabbit-keeper.

However, as a concrete suggestion, a good number to start with is a minimum of ten does and one buck. This number will give you more pleasure to work with and will cost less, possibly, than a trio of registered animals. You should, nevertheless, concentrate on eligible-for-registration stock, if you plan to sell breeders.

Goal #6
Use Proper Feed and Feeding Techniques

In the days when the rabbit industry was young, the feeds that were most desirable were quite different from those used today. Oats and alfalfa was first choice.

Both cultivated and wild greens were considered natural, nutritious, inexpensive food for rabbits. Since they are rich in proteins, minerals and vitamins some authorities maintain that a mature rabbit can be kept in good health indefinitely on greens alone.

It has been recommended that if you have access to greens in your garden then you should utilize them to the best advantage, because they are palatable for rabbits and lower the cost of feeding them. However, if your rabbitry contains several hundred animals the labour involved is tremendous.

Simplicity in feeding should be favoured for it reduces the time and thus cuts labour costs if a sizeable rabbitry is in operation.

For many years rabbits were fed a combination of dry grains which included wheat, corn, oats, and barley. Timothy, alfalfa and clover hay was fed in addition to these grains.

This mixture has been used for years until recently an advance, and a great advantage in feed for rabbits has been manufactured by feed concerns in the form of rabbit pellets.

Today these rabbit pellets are found to be superior to other types of feed and have eliminated the use of greens and vegetables, as well as hay, as such, by many rabbit keepers.

Actually, when pellets are fed to the herd, all the various grains, plus, are included in the mixture giving the rabbits a well-balanced diet.

The writer endorses the use of good clean alfalfa, timothy or clover hay in addition to the pellets as a supplement which supplies roughage for the animals. The hay fed should be given in small quantity so that none will be left over to become soiled or mildewed.

Pellets are produced by many manufacturing firms and sold under various trade names. By using the pellet method of feeding rabbits, much valuable time is saved and overhead costs are lowered.

It pays to shop around for the best deal when purchasing pellets. Do not be content to buy the cheapest if they give inferior results. Once you get satisfactory pellets by all means stick to them until you can prove others to be better.

When you get in a new stock of pellets do not pile them on what is already on hand. Using feed in strict rotation as it is purchased keeps fresh food available for the rabbits all the time as it should be.

Automatic feed hoppers placed on the outside of the cage is the obvious choice because it cuts out the risk of rabbits standing in the food containers thus fouling the pellets and wasting them.

The hoppers can be filled from outside in less time than it takes to open doors and fill inside containers, thus reducing the daily chore of feeding.

Your feeding programme should be set and strictly adhered to. Not often are specific directions supplied with the product, therefore you should set up your own schedule.

Proper feeding is very essential if your venture is going to be worthwhile. Do not overfeed and at the same time remember that under-feeding is far from being desirable.

It has been proven by leading commercial operators and at experimental stations where rabbits are studied, that once-a-day feeding is satisfactory.

This can take place either in the morning or evening which ever time suits the rabbit keeper, however, because rabbits by nature eat more at night it is recommended that evening feeding take preference in this regard. Which ever time you decide on, you should definitely aim at feeding the rabbits the same hour each day. Whether you feed once, twice or more times a day, regularity is most important.

The amount of daily ration for rabbits will depend on several aspects such as, the breed to which it belongs, the age, (i.e.) whether young or mature, dry does for maintenance or does with litters for growth.

The climate also has a bearing on the consumption of food by rabbits. Usually in cold climates rabbits eat more. The size of the cage is another factor, for rabbits who can exercise will eat more than those in crowded quarters.

Generally speaking, a doe or buck in regular service consumes 4 to 5 ounces of pellets per day. Size, weight and activity of a rabbit are influential factors determining the need for more or less feed. One week before kindling, feed the doe all she wants to eat. Does with litters should be on full feed with pellets in the hoppers continually. Weaned young should be kept on full feed also. Young replacement stock requires plenty of body development. By the time replacement does reach five months of age they should be watched so they won't get too fat, which makes it difficult to breed them.

A successful programme for feeding domestic rabbits include rations that will be economical yet satisfactory. Pellets are suitable for this purpose if properly fed. They assist in maintaining a high natural resistance to disease and aid in producing maximum growth.

Goal #7
Plan Breeding Schedule

In order to develop a good breeding schedule in correct breeding procedure, many lessons must be learned relative to the production in rabbits.

This knowledge will enable you to follow good breeding practices which in the long run will be more beneficial than just taking things for granted.

A most important fact for you to remember is that the doe must be healthy in order to produce. In fact the selection of both does and bucks for breeding purposes must be exercised with care, and not only must the health of the animals be taken into consideration, the age of the breeders must also be taken into account.

The number of litters which a doe should be allowed to have during a given time is often the subject of argument for many rabbit keepers.

Many fanciers prefer to breed a doe twice and occasionally three times a year.

Some commercial breeders are content with four litters a year while others will breed for five or even six kindling in the same period.

Quoting from an article in a recent issue of "Rabbits in Canada," a rabbitry in British Columbia reports: "We are on a fast breeding programme: barring no misses, we have eight litters per year per doe. We have found the does much easier to breed all year round on this tighter breeding programme,"

There is little harm likely to result from frequent breeding. Indeed the reverse can be true providing of course that feeding and management are satisfactorily executed. Frequent breeding under these circumstances is to be desired.

A doe that is bred under 7.5 lbs. at age 5 months may always stay small, although this is debatable, nevertheless a doe that is bred under these conditions will lack physical stamina. Her productive life may be damaged and shortened.

It is wise for the novice to adopt a breeding schedule that can easily be maintained.

The proper age for breeding in most breeds to be fully developed is 8 months. Breeds for commercial fryer production are fully developed at 6 months. Of course, judgement should be used by the beginner in this matter which will come by experience.

Bucks can be put into service at 5 1/2 to 6 months if they are in good physical condition. For continuous good results rabbits of any type should not be bred unless they are extremely well developed.

For the beginner a good schedule to follow until, through experience he wishes to make a change, is as follows: Breed the doe 56 days (8 weeks) after she has kindled, at which time the young from the previous litter are weaned. Under this system four litters per year are produced.

Later on you could adopt the 6 week breeding schedule and breed for five litters per doe per year. The litter would still remain with the mother for a full 56 days at which time they are weaned, and this allows 17 days of rest before she kindles again.

It is known that rabbits have a 16-day heat cycle and are fertile at all times except for two days at the end of a cycle and two days at the beginning of the next. This means they have twelve days fertile, and four days infertile, and these periods are recurring. Therefore, a doe may be bred and can conceive at any time during the heat period.

The doe shows signs of heat by becoming restless, rubbing her chin on the cage or other objects, attempting to join other rabbits and so on. The lack of interest in the buck is a sign she is not in heat.

The period of time which elapses between mating and kindling is known as the gestation period. Domestic rabbits have a gestation period of thirty-one days. This time period occasionally becomes shorter or longer by a day or two.

The most favourable time for breeding the does is early morning or in the evening, and it should take place before the rabbits are fed.

The commercial operator has to find time throughout the year to keep the breeding on a regular schedule. He must carry out this chore every day or every week depending on the size of his operation.

When breeding a doe, take her to the buck's cage and watch until he has serviced her. Never take the buck to the cage of the doe. This will cause waste of time while the buck is getting acquainted with his new surrounding and the doe may resent the intruder.

Mating usually takes place within a few minutes. Service is complete when the buck falls on his back or off to his side. As soon as the doe accepts service she should be returned to her own cage immediately.

If a doe will not accept service she should be taken back to her cage, and again tried the following day, or days until she does. It is sometimes advisable to "test breed" the doe by taking her back to the buck in about three days; if she is pregnant she will fight him. Do not forget to keep precise records as to which doe is bred to which buck.

Goal #8
Exercise Care of Litters

To effectively manage a rabbitry many practices must be carried out each and every day of the year. All details must be faithfully and regularly attended to. The rabbit keeper who is experienced and successful knows what to do, how to do it and when it should be done.

The novice must learn that successful management demands attention to all details, however, management requires practice. Rabbit keeping systems differ, and what may be considered good management practices for some may be discarded by others.

The majority of rabbit breeders are anxious to assist newcomers to start on the right foot, and because this goal will probably be the most interesting to the novice, the writer proposes to aid in this direction.

Preparing for the young is an important period in raising rabbits which should receive close attention by the owner. Your successful operation will depend largely on your observance and diligence in making preparation for the litters as they become due.

Once the does are mated they do not require any special care. Sometimes a doe will start to prepare her nest 17 to 20 days after mating. This is known as false pregnancy. When this happens the doe should be bred again immediately.

Three or four days before the doe is due to kindle, place a nest box in the cage. It is important to prepare the nest box with care. A good method to use is to cover the bottom of the box with a half inch of good dry sawdust. While some breeders advocate avoiding sawdust in nest boxes because the fine grains may cause nasal problems and stick to the breasts of the nursing doe, making the young ill, the writer has found sawdust to be no problem in this regard. It will help to soak up the urine, keeping the nest dry, and will also make the cleaning of the box much easier because it prevents the straw and other material from sticking to the bottom.

Place a good bedding of soft straw on top of the sawdust, preferably oats straw, about three quarters full during the colder weather, while less for the warmer weather will suffice.

From this point on the doe will prepare the nest herself. She will pluck mouthfuls of fur from her body to mix with the straw and line a comfortable nest for her young.

A day or so before kindling the doe may go off her feed, not eating as much as usual. This should cause no alarm but rather, it should be regarded as a good sign that she is nearing the kindling stage. Try feeding her a small quantity of green feed or a piece of dry bread but do not disturb her more than necessary.

Do not allow any visitors or cause any noise to disturb the process of kindling. Does expecting to kindle should be watched occasionally, especially in the early morning and late at night.

BREEDING & KINDLING CHART

Instructions: If the doe is bred on specific date on the top line, her kindling date will be immediately below the breeding date on the second line.

Month	1	2	3	4	5	6	7	8	9	10	11	12	13	14	15	16	17	18	19	20	21	22	23	24	25	26	27	28	29	30	31
JANUARY	1	2	3	4	5	6	7	8	9	10	11	12	13	14	15	16	17	18	19	20	21	22	23	24	25	26	27	28	29	30	31
FEBRUARY	1	2	3	4	5	6	7	8	9	10	11	12	13	14	15	16	17	18	19	20	21	22	23	24	25	26	27	28	1	2	3
FEBRUARY	1	2	3	4	5	6	7	8	9	10	11	12	13	14	15	16	17	18	19	20	21	22	23	24	25	26	27	28			
MARCH	4	5	6	7	8	9	10	11	12	13	14	15	16	17	18	19	20	21	22	23	24	25	26	27	28	29	30	31			
MARCH	1	2	3	4	5	6	7	8	9	10	11	12	13	14	15	16	17	18	19	20	21	22	23	24	25	26	27	28	29	30	31
APRIL	1	2	3	4	5	6	7	8	9	10	11	12	13	14	15	16	17	18	19	20	21	22	23	24	25	26	27	28	29	30	1
APRIL	1	2	3	4	5	6	7	8	9	10	11	12	13	14	15	16	17	18	19	20	21	22	23	24	25	26	27	28	29	30	
MAY	2	3	4	5	6	7	8	9	10	11	12	13	14	15	16	17	18	19	20	21	22	23	24	25	26	27	28	29	30	31	
MAY	1	2	3	4	5	6	7	8	9	10	11	12	13	14	15	16	17	18	19	20	21	22	23	24	25	26	27	28	29	30	31
JUNE	1	2	3	4	5	6	7	8	9	10	11	12	13	14	15	16	17	18	19	20	21	22	23	24	25	26	27	28	29	30	1
JUNE	1	2	3	4	5	6	7	8	9	10	11	12	13	14	15	16	17	18	19	20	21	22	23	24	25	26	27	28	29	30	
JULY	2	3	4	5	6	7	8	9	10	11	12	13	14	15	16	17	18	19	20	21	22	23	24	25	26	27	28	29	30	31	
JULY	1	2	3	4	5	6	7	8	9	10	11	12	13	14	15	16	17	18	19	20	21	22	23	24	25	26	27	28	29	30	31
AUGUST	1	2	3	4	5	6	7	8	9	10	11	12	13	14	15	16	17	18	19	20	21	22	23	24	25	26	27	28	29	30	31
AUGUST	1	2	3	4	5	6	7	8	9	10	11	12	13	14	15	16	17	18	19	20	21	22	23	24	25	26	27	28	29	30	31
SEPTEMBER	1	2	3	4	5	6	7	8	9	10	11	12	13	14	15	16	17	18	19	20	21	22	23	24	25	26	27	28	29	30	1
SEPTEMBER	1	2	3	4	5	6	7	8	9	10	11	12	13	14	15	16	17	18	19	20	21	22	23	24	25	26	27	28	29	30	
OCTOBER	2	3	4	5	6	7	8	9	10	11	12	13	14	15	16	17	18	19	20	21	22	23	24	25	26	27	28	29	30	31	
OCTOBER	1	2	3	4	5	6	7	8	9	10	11	12	13	14	15	16	17	18	19	20	21	22	23	24	25	26	27	28	29	30	31
NOVEMBER	1	2	3	4	5	6	7	8	9	10	11	12	13	14	15	16	17	18	19	20	21	22	23	24	25	26	27	28	29	30	1
NOVEMBER	1	2	3	4	5	6	7	8	9	10	11	12	13	14	15	16	17	18	19	20	21	22	23	24	25	26	27	28	29	30	
DECEMBER	2	3	4	5	6	7	8	9	10	11	12	13	14	15	16	17	18	19	20	21	22	23	24	25	26	27	28	29	30	31	
DECEMBER	1	2	3	4	5	6	7	8	9	10	11	12	13	14	15	16	17	18	19	20	21	22	23	24	25	26	27	28	29	30	31
JANUARY	1	2	3	4	5	6	7	8	9	10	11	12	13	14	15	16	17	18	19	20	21	22	23	24	25	26	27	28	29	30	31

It is a good idea to take a quick last look at all rabbits in the rabbitry before retiring. Some does may kindle on the wire and if you are on guard many young may be saved that otherwise would die. Sometimes the entire litter may be lost when the doe kindles on the floor of the cage, especially during the cold weather.

The following breeds are pictured above from left to right: Giant Chinchilla, Middle Californian & A Checkered Giant All Breed Champions at the Canadian Agricultural Royal Winter Fair in Toronto, Canada
Above Rabbits owned by
Diamond J. Rabbitry, Brighton, Ontario, Canada

A day or so after kindling takes place, inspect the litter and remove any dead or deformed young. This should be done with care. The doe may be given a treat of tempting food to hold her attention while the examination is in progress. It is also a good idea to talk to the mother rabbit to let her know that a stranger is not disturbing her babies.

Usually there are six, eight or more young in the nest. It is not unusual to find as many as ten, twelve or even fourteen after a doe has kindled. It has been reported that twenty-three rabbits have been found in one litter. This number, of course, is the exceptional.

It has been found practical to reduce the litter to eight of the strongest animals and foster out the rest to another doe that has produced a smaller litter about the same time. This has been done successfully on many occasions as experienced rabbit keepers will testify. A little Vicks placed on the nose of the mother rabbit will help in this regard. Sometimes the sprinkling of a small amount of talcum powder around the nest will help to reduce the foreign odour of the newcomers.

After you have inspected the nest box and made the necessary changes, if necessary, you should leave the doe and litter undisturbed but, frequent checks are absolutely necessary to maintain a healthy nest box for the young.

It takes ten to twelve days, after birth, for the young to open their eyes. Upon inspection, if the eyes do not respond and open naturally, it may be necessary to aid with the opening artificially. This can be done by bathing the eye lids with a solution of boric-acid and warm water. It will help to prevent blindness of the animal.

When the young are twenty-one days old they will come out of the nest box and begin to eat solid food. Sometimes one or two may jump out in search of food from the mother and may have to be helped back to the nest. Rabbits do not help their young as is the custom for cats.

The nest box should be removed when the young are four weeks old, cleaned, sterilized and stored in readiness for use.

The young should remain with the mother for eight weeks, if possible, at which time they should be weaned, sexed, separated or shipped to market depending for which purpose they are intended to be used.

The rabbit should never be lifted by the ears. It may cause permanent damage. Use one hand to grasp the loose skin at the back of the neck and the other hand to support the weight placed under the rump. Fryers may be lifted comfortably by the loin.

Protecting the rabbits from extreme cold weather, keeping them dry and having proper temperature control in the rabbitry is very important to a successful business of domestic rabbit raising.

Goal #9
Control Diseases

In controlling diseases of domestic rabbits it should be strongly emphasized that, "Prevention is Better than Cure."

The best assurance against any form of sickness or disease is to have all cages, nest boxes and equipment in the rabbitry kept in a sanitary condition.

The problem of disease in rabbits is best solved by forestalling it through precautions well in advance.

Disease in rabbits can very often be attributed to anything but good husbandry. The most common circumstances contributing to disease in any rabbitry are dirty feeders, dirty drinking containers, dirty cages, and over-crowding which, summed up, means faulty management.

A rabbit keeper should guard against, and prevent these conditions from encouraging diseases in his herd, which, if not checked, may cause serious trouble. Proper sanitary measures will assist in maintaining a herd in a healthy condition, which every owner should strive for.

Disease can be spread by using the same cleaning utensils, such as a brush or cloth, to clean feeding and watering equipment from one cage to another. This practice should be avoided.

The nest boxes should be cleaned with some disinfecting agency and set out in the sun or burned out with a gas blow torch before using again.

Cages and feeding utensils should be cleaned regularly. Do not let the food and drinking water get contaminated. Give the rabbits food and drinking water that is fresh and clean each day.

Before restocking with healthy rabbits or bringing other rabbits into the rabbitry all cages for this purpose should be thoroughly clean and disinfected. Each time a rabbit is moved to another cage it should be placed in a clean one. This will prevent the spread of disease if there is any present.

To control flies, which is by no means welcome in any rabbitry, dispose of any dead rabbits and infected materials immediately, by burying it in deep soil or disposing of it by burning. Flies can spread disease within the rabbitry or introduce it from an outside source.

Proper ventilation in the rabbitry is necessary. Dampness, temperature change or poor circulation of air lower the vitality of the animals and increase their susceptibility to disease.

All sick animals should be isolated and sometimes it is far better to have them destroyed. For instance, if sick rabbits do not respond to treatment it is better to dispose of them to prevent the spread of disease to the other rabbits in the herd.

It is not the purpose here to give a description of, or a cure for, the many rabbit diseases but only to suggest means of preventing any disease from entering your rabbitry. The diagnosis and treatment should be left to a qualified veterinarian in case of an ailment beyond the control of the rabbit keeper.

Some of the most common diseases prevalent in rabbits are:
Colds, Pneumonia, Vent Disease, Abscesses, Droopy Ears, Tapeworm, Sore Hocks, Coccidiosis, Ear Mange, Buck Teeth, Mucoid Enteritis and Wry Neck.

Goal #10
Sell Production at a Profit

It is not the intention of the writer to give an account in detail as to how your production should be prepared to be sold at a profit, but rather to partially enumerate the items that can be sold.

Except for those who raise rabbits for the family tables, for a hobby, or for pets, the prime purpose of the business is to make a profit.

If you concentrate your efforts on good management and to increasing the size and productivity of your rabbit venture your profit will increase.

Taking into consideration the stock and equipment which is available today for the interested people, the rabbit meat production can offer a good profit from capital invested.

By aiming at the goals outlined in the preceding pages of this book you stand a very good chance of reaping a profit from your rabbit business. Start with cheap breeding stock, inadequate housing, ill-constructed nest boxes, etc., and you will only be disillusioned.

Keep in mind that the most important thing to realize is that you are running the business. This enterprise will increase or decrease according to the profit realized.

The profit you will get from the sale of your rabbits and rabbit products or by-products, will by and large depend upon the market and the prevailing price to be paid, and this price is usually governed entirely by the demand.

Prices paid will naturally be in accordance with prices received. The profit to the producer will be the difference between the production cost and the cash intake. The producer should therefore concentrate on producing the market weight of rabbits in the shortest possible time.

There are, besides meat, more products and by-products to be marketed from a rabbitry. Sales of livestock for breeding purposes, for exhibition and for pets should be established. However, meat is the prime purpose for the commercial operator.

A considerable amount of meat is sold to meat shops, hotels, hospitals, families and so on, by the smaller rabbit keeper who may realize a sizeable profit from these sales.

Those who are keeping a sizeable herd could sell some of his products locally but it almost becomes a necessity to sell his entire production of meat to a wholesale processor.

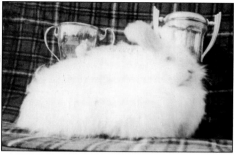

A large number of furs are used by the fur trade and although the price is small, a certain amount of the cash can be realized from the sale of this product. The market is not quite dependable but if the pelts are cared for and stored for awhile the market could improve and then the sale could be made.

To sell rabbits for breeding stock you should have the best possible pedigreed and registered stock available. If you can produce first class breeding stock then by all means sell them for this purpose, for they demand a higher price than you would otherwise get for second or third grade animals.

Champion Angora Rabbits
Diamond J. Rabbitry, Brighton, Ontario, Canada

Certain types of work is done at laboratories and the domestic rabbit may be used for this purpose. For example, the rabbit is used for testing drugs and for general research. The demand is rather variable and sometimes the price is not encouraging. However, much higher prices are paid for quality animals with special characteristics.

Earthworms raised under cages is a wonderful moneymaking combination. Rabbit manure and waste feed around the cages is the best food for earthworms to be found anywhere. In addition to your profit from rabbits, you can raise earthworms under the cages at no extra cost, no feed to purchase and practically no work. In this way you can earn extra income.

You can turn rabbit manure into money. A rabbit will produce a good quantity of manure in the course of time, therefore you should not overlook the profit possibilities to be gained from this by-product.

In conclusion, remember that rabbits offer unusual benefits to anyone interested in pursuing the domestic rabbit raising business, whether it be for a delightful hobby, a family project or a profitable commercial business.

Domestic Rabbits

American	Himalayan
Angora	Lilac
Belgian Hare	Lop
Beveren	Netherland Dwarf
Californian	New Zealand
Champagne D'Argent	Palomino
Creme D'Argent	Polish
Checkered Giant	Rex
Chinchilla	Sable
Dutch	Satin
English Spot	Siamese Sable
Flemish Giant	Silver
Florida White	Silver Fox
Havana	Silver Martin
Harlequin	Tan

Recipes
for Preparation of Domestic Rabbit Meat

The meat of the domestic Rabbit is white, fine-grained and delicately flavoured. It may be prepared as you would chicken. Young animals may be roasted or fried, while older animals should be braised or cooked in liquid. It is an excellent source of high-quality protein and has little bone and fat content. Many tasty dishes can be prepared with rabbit meat for the family table. There are many ways to prepare this delicious tender meat. The following are a few recipes to show how this meat can be prepared:

Stewed Rabbit

rabbit cut into serving pieces (about 4 lbs.)
1 1/2 teaspoons salt
hot water
1 chopped onion

Put rabbit into pan large enough to hold the pieces without crowding. Add salt and enough water to half cover the rabbit. Cover pan and cook on low heat for 1 1/2 hours, or until the meat is very tender. Add more water during cooking if necessary. Serve hot with gravy made by thickening the broth and seasoning as desired. Makes approx. 8 servings.

Hassenpfeffer

2 rabbits, cut in serving pieces
1 tablespoon salt
2 medium onions, sliced
3 whole cloves
1 bay leaf
6 whole allspice

1/2 cup white vinegar
water
2 tablespoons butter
2 tablespoons flour
pepper

Place the pieces of rabbit in a casserole dish and sprinkle with salt. Heat, but do not boil, onions, cloves, bay leaf, and allspice in vinegar and pour this over the rabbit. Let cool and set in refrigerator for two days, turn the pieces occasionally. Put in a Dutch oven with the marinade, adding enough water to cover. Cover the pan and simmer until rabbit is tender (1-1 1/2 hours). Melt the butter and stir in flour. Add strained rabbit broth and cook for 5 minutes, stirring constantly. Pour over rabbits. Serve with potato balls fried to a golden brown colour and cole slaw. Makes 6-8 servings.

Roast Rabbit

A rabbit ideal for roasting weighs between 3 to 5 lbs. dressed. The whole rabbit (not cut up) is used for this purpose. After washing thoroughly inside and out, fill the rabbit with your favourite dressing and sew up the carcass. Put strips of bacon across the rabbit, or pour melted margarine or butter over the rabbit. Bake slowly for 2 to 3 hours depending on the age of the rabbit. Serve with brown gravy from the broth, season to taste. For gravy, use 2 tablespoons of pan drippings and 1-1 1/2 tablespoons of flour. Blend drippings and flour thoroughly in the skillet. Add one cup of milk and stir until it thickens as desired.

Additional Sources of Information

1. American Rabbit Breeders Association
    Send all official questions concerning the ARBA to:

    Glen Carr
    c/o ARBA
    PO Box 426
    Bloomington, IL 71702
    Tel: (309) 664-7500
    Fax: (309) 664-0941

    ARBA Web Page: http://www.arba.net

2. National Angora Rabbit Breeders Assoc.

    Vicki Johnston
    c/o NARBA - Secretary
    2380 Co Rd NE
    Nelson, MN 56355
    Tel: (320) 762-0376